FUTURE SOURCES

© Aladdin Books Ltd

Designed and produced by
Aladdin Books Ltd
70 Old Compton St
London W1

First published in
Great Britain in 1985 by
Franklin Watts
12a Golden Square
London W1

ISBN 086313 310X

Printed in Belgium

FUTURE SOURCES

JAMES STRACHAN

Illustrated by
Ron Hayward Associates

Consultant
Stewart Boyle

Franklin Watts / Aladdin Books
London : New York : Toronto : Sydney

F
Introduction

Imagine what it would be like if you had no gas or electricity to heat and light your house. Imagine that there was no petrol to fuel the cars, buses and trucks that transport people and goods. Factories would have no power to run their machines, so there would be nothing in the shops to sell. Our way of life depends on having a constant source of power and fuel.

In this book you will find out about some of the ways people are planning to provide us with energy sources for the future.

Energy consumption: Hong Kong at night

Contents

Energy today

In the past 50 years, we have used up over 40 per cent of the world's known fossil fuel deposits – those of oil, coal, and gas. At the present rate of use, oil and gas will run out by the year 2050. Coal has a longer lifespan – with new technologies such as the computer-controlled drill shown on the right, our coal supplies will last for at least 250 years. But coal is not a clean fuel – it causes pollution, and so we also need to develop other technologies which will reduce this pollution.

Pollution from coal burning

In poorer countries, governments cannot afford to spend large sums of money on huge new energy projects. The cost of building just one nuclear power station, for example, runs into hundreds of millions of pounds. But traditional energy sources, such as firewood, are becoming scarcer and scarcer, and many people have to spend as much on fuel as they do on food. New, cheaper, sources of energy are needed both in the developing countries and in the West.

New mining methods

Everyday fuels

In the poorest countries of Africa, firewood can make up a total of over 80 per cent of total energy usage. In many of these countries, the cutting of trees and shrubs for firewood has made once fertile land increasingly arid and difficult to farm. As firewood becomes scarce, the problems can only get worse, and so urgent solutions are needed.

In India, animal dung is dried and formed into bricks to be used as fuel. And in many parts of China animal and human wastes are collected and stored. After a period these give off methane gas which can be used for cooking and heating. Another solution is to use the waste material from plant crops for fuel.

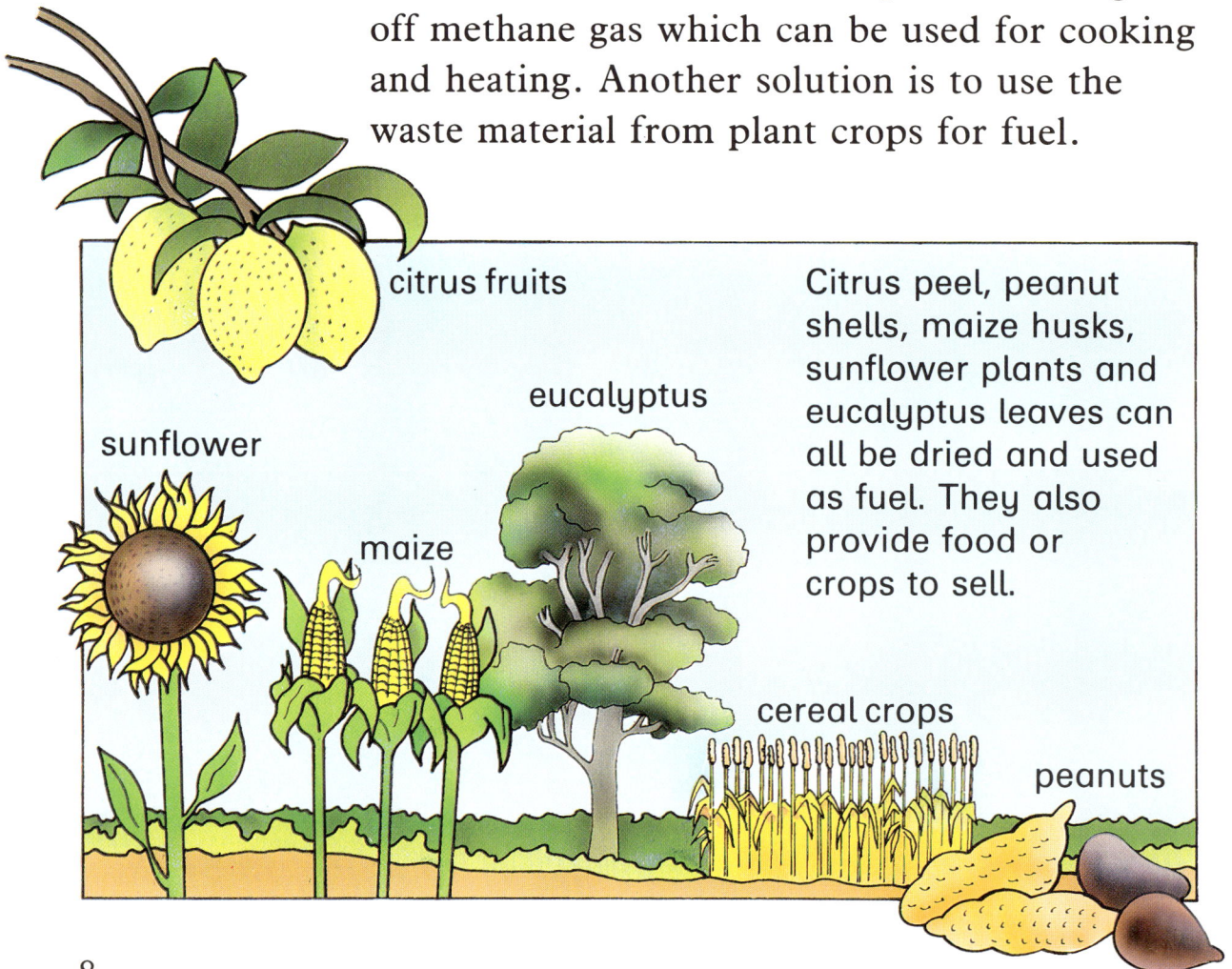

citrus fruits

sunflower

eucalyptus

maize

cereal crops

peanuts

Citrus peel, peanut shells, maize husks, sunflower plants and eucalyptus leaves can all be dried and used as fuel. They also provide food or crops to sell.

Stockpiling cow-dung in India

9

Save it

It is not just coal, oil and gas that the Earth is running out of – metals such as tin, copper and aluminium are also becoming scarcer. Yet these materials are found in the rubbish we carelessly throw away every day.

This rubbish can be used to help save our resources. Metals needed by industry are separated out, to be used again. Newspapers and food wrappings can be recycled or burned in electricity power stations. Some small plants are already in operation.

Separation plant . . .

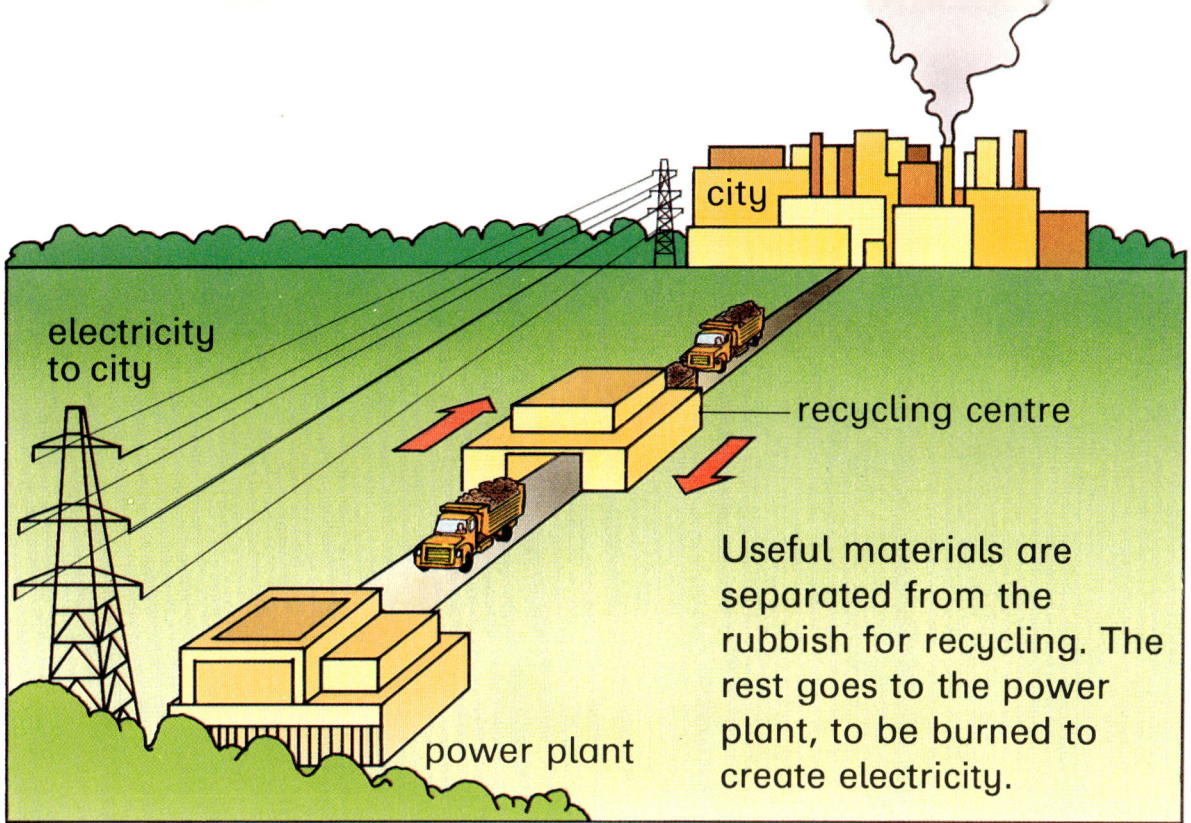

city

electricity
to city

recycling centre

Useful materials are
separated from the
rubbish for recycling. The
rest goes to the power
plant, to be burned to
create electricity.

power plant

. . . power plant

Green power

In Brazil, most of the cars now run on "alcool" made from crushing specially grown plants. But crops raised for fuel take up valuable land needed to grow food for the world's poor.

The kelp farm off the west coast of the USA, shown opposite, doesn't have this drawback. The giant kelp grows at a rate of 60 cm (24 ins) a day, and is harvested three times a year. The kelp is processed to provide chemicals, animal feed, and methane gas fuel. It is estimated that a kelp farm covering about 850 sq km (330 sq miles) could supply the whole of the USA's gas needs!

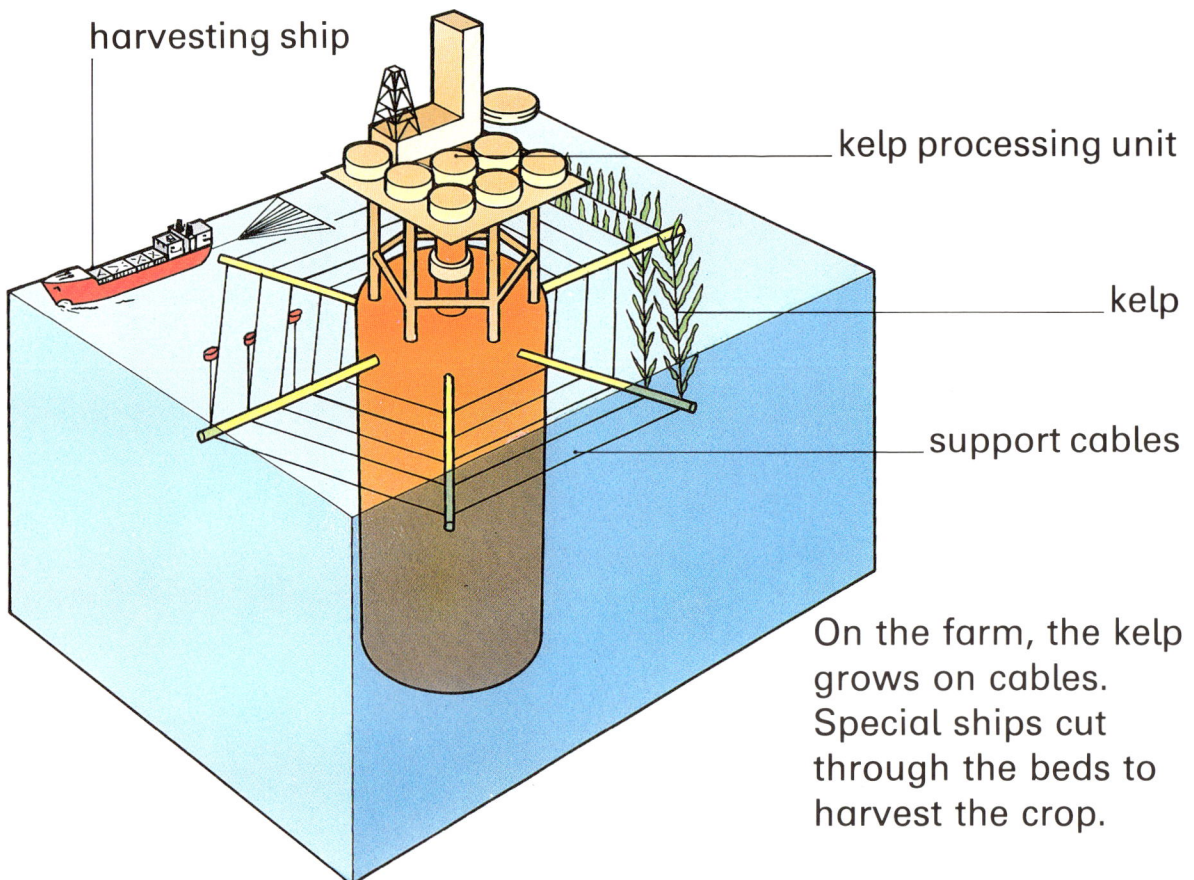

harvesting ship

kelp processing unit

kelp

support cables

On the farm, the kelp grows on cables. Special ships cut through the beds to harvest the crop.

The fuel cell

Engineers are experimenting with a new way of making electricity – the fuel cell. The fuel cell is like a battery, only one which will never run out. In it hydrogen and oxygen gases react with special chemicals to produce electricity. The hydrogen and oxygen combine to make water. The water can be treated so that the gases can be separated and used in the fuel cell again.

Small fuel cells have been used to power systems on board the space shuttle, and even these are still very expensive to build. But engineers one day hope to build cells that can supply whole city areas with electricity. ·

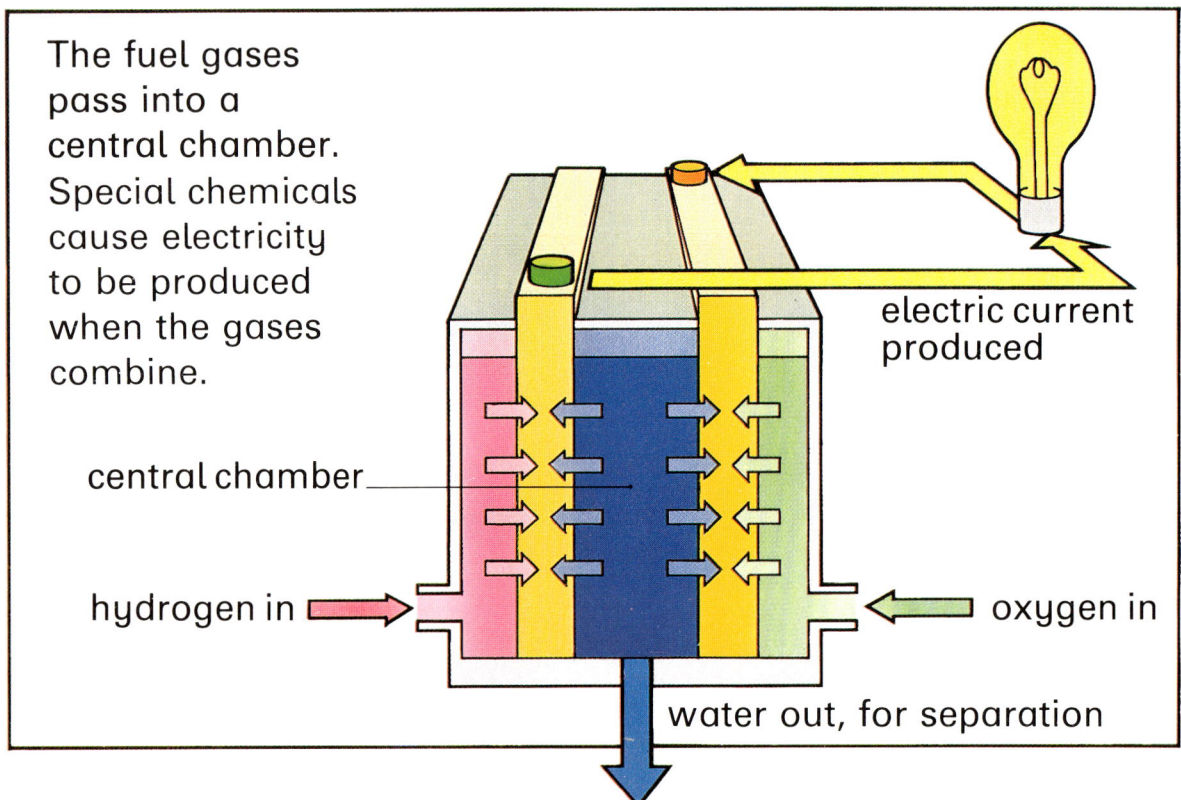

The fuel gases pass into a central chamber. Special chemicals cause electricity to be produced when the gases combine.

electric current produced

central chamber

hydrogen in

oxygen in

water out, for separation

Space shuttle launch

Solar power tower,
Mojave Desert, California

Solar energy

Solar power – power from the sun – is being used in projects worldwide. It can be used to heat water in the home, and so reduce demands on electricity or gas. In irrigation schemes, pumps powered by solar electric cells deliver water to crops.

But solar power can also generate electricity on a larger scale. Shown opposite is a solar "power tower" in the USA. Over 1,800 mirrors direct the sun's heat to a receiver at the top of the tower. This concentrated heat produces steam to drive turbines which create enough electricity for about 2,500 homes.

Solar pump in the Philippines

Wind and waves

The wind and waves have enormous energy. The wind's energy has been used for centuries, and many small wind power projects now generate electricity for farms and villages.

Taming the sea is much harder. This is because a structure able to withstand storm conditions at sea would be expensive and difficult to build. But several experimental ideas are being studied. One such is the Breakwater scheme shown below, designed to be placed off the west coast of Scotland and in Norwegian fiords.

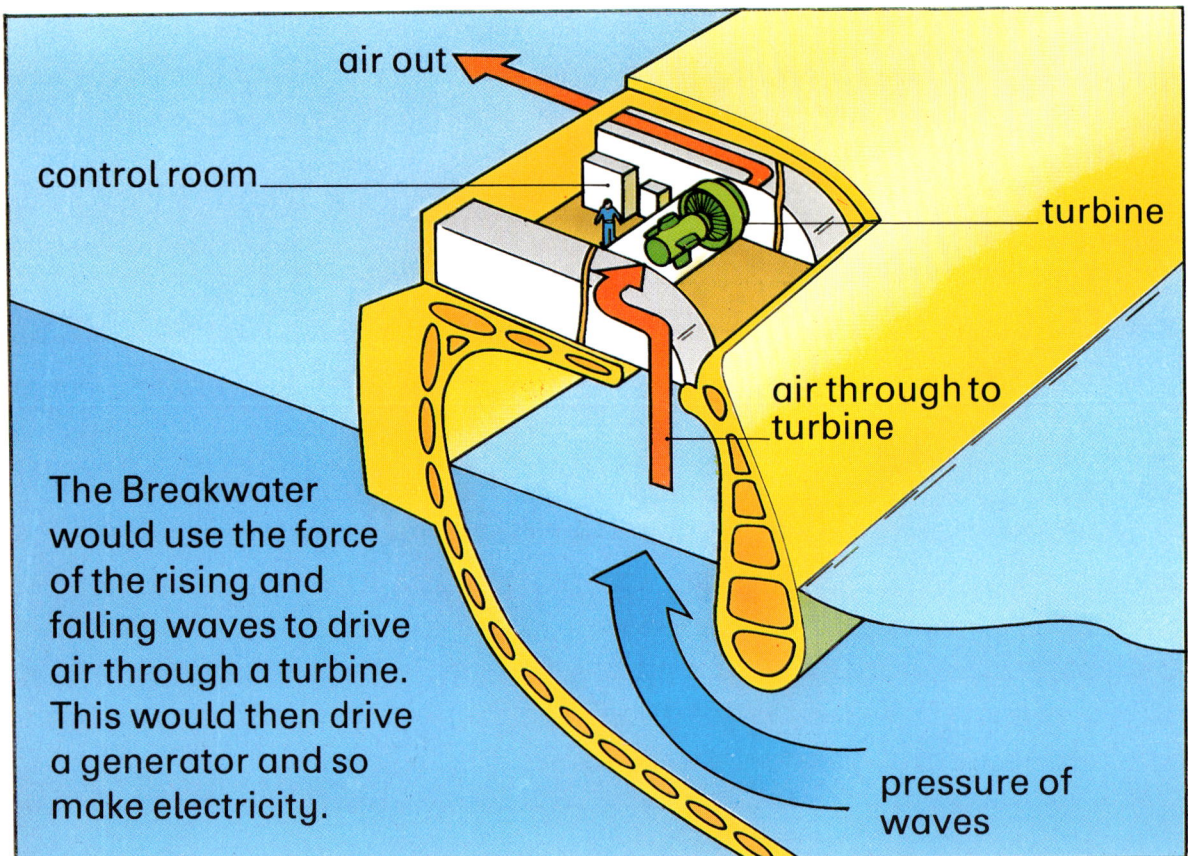

air out

control room

turbine

air through to turbine

The Breakwater would use the force of the rising and falling waves to drive air through a turbine. This would then drive a generator and so make electricity.

pressure of waves

To obtain much larger amounts of electricity from the wind, giant "wind farms" have been proposed, such as the one shown here. These could be sited in shallow offshore waters, where winds are strongest, and their presence would not spoil country landscapes.

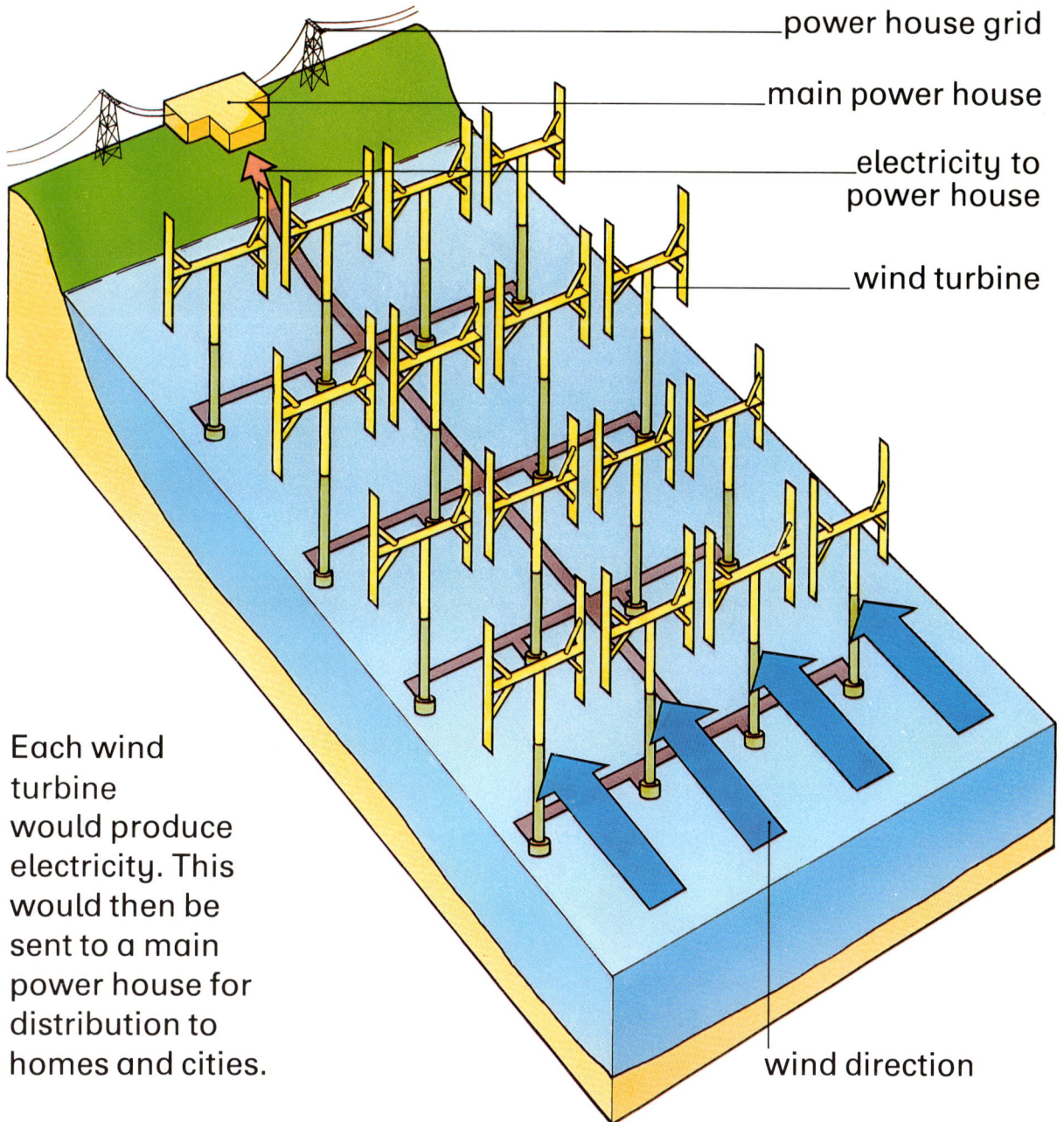

power house grid

main power house

electricity to power house

wind turbine

Each wind turbine would produce electricity. This would then be sent to a main power house for distribution to homes and cities.

wind direction

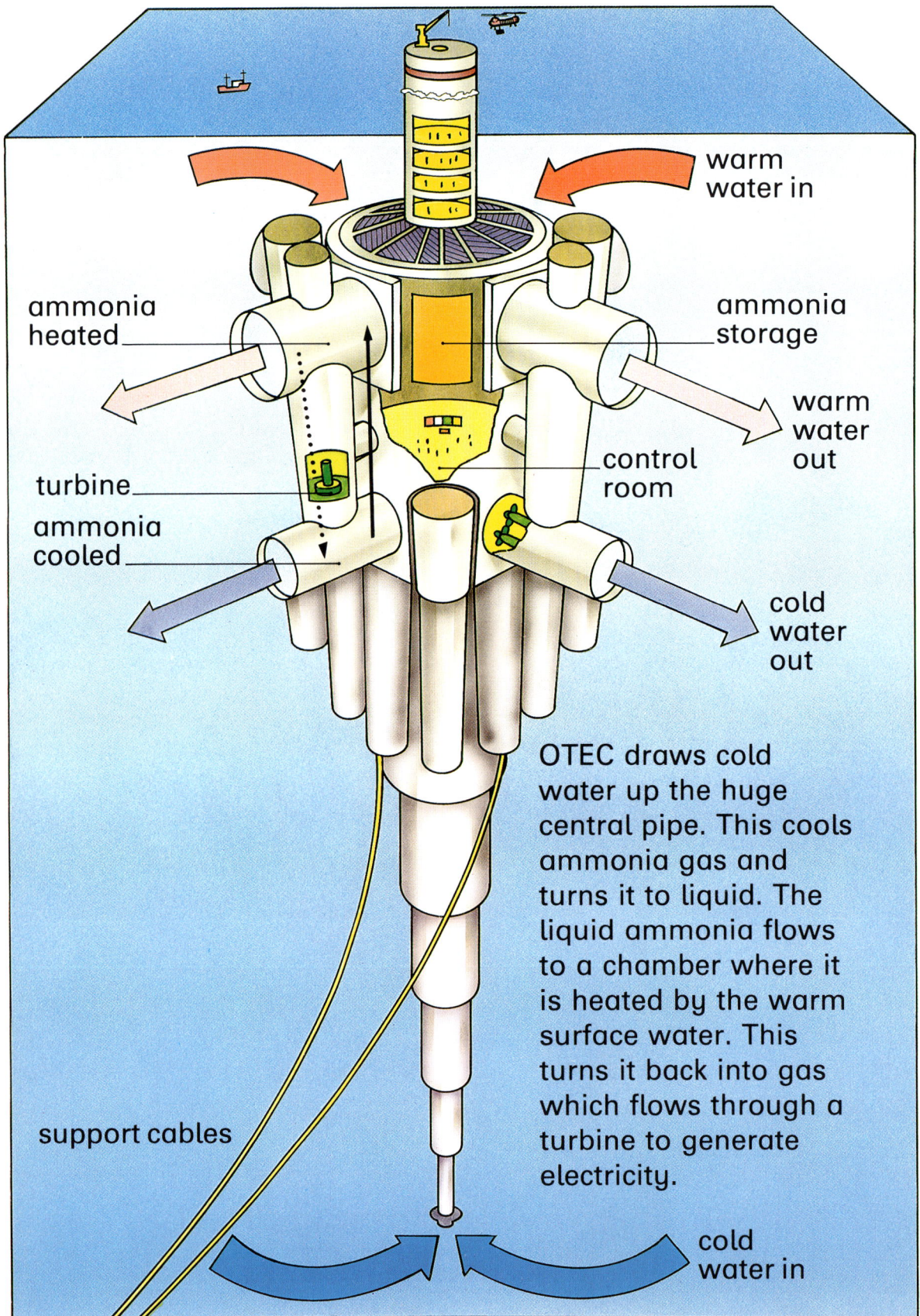

warm
water in

ammonia
heated

ammonia
storage

warm
water
out

turbine

control
room

ammonia
cooled

cold
water
out

OTEC draws cold water up the huge central pipe. This cools ammonia gas and turns it to liquid. The liquid ammonia flows to a chamber where it is heated by the warm surface water. This turns it back into gas which flows through a turbine to generate electricity.

support cables

cold
water in

Ocean giants

The massive structure shown on the opposite page is called an "Ocean Thermal Energy Converter" – OTEC, for short. It is designed to generate electricity in tropical waters, where the sea's surface temperature is much higher than it is at depths of 1,000 m (3,280 ft) or more. In colder climates, the temperature difference is less, and the system would be less effective.

Already, two small experimental OTECs have been built and tested off Hawaii in the Pacific Ocean. The United States hopes to have giant plants working early next century, each generating enough electricity to supply a large town.

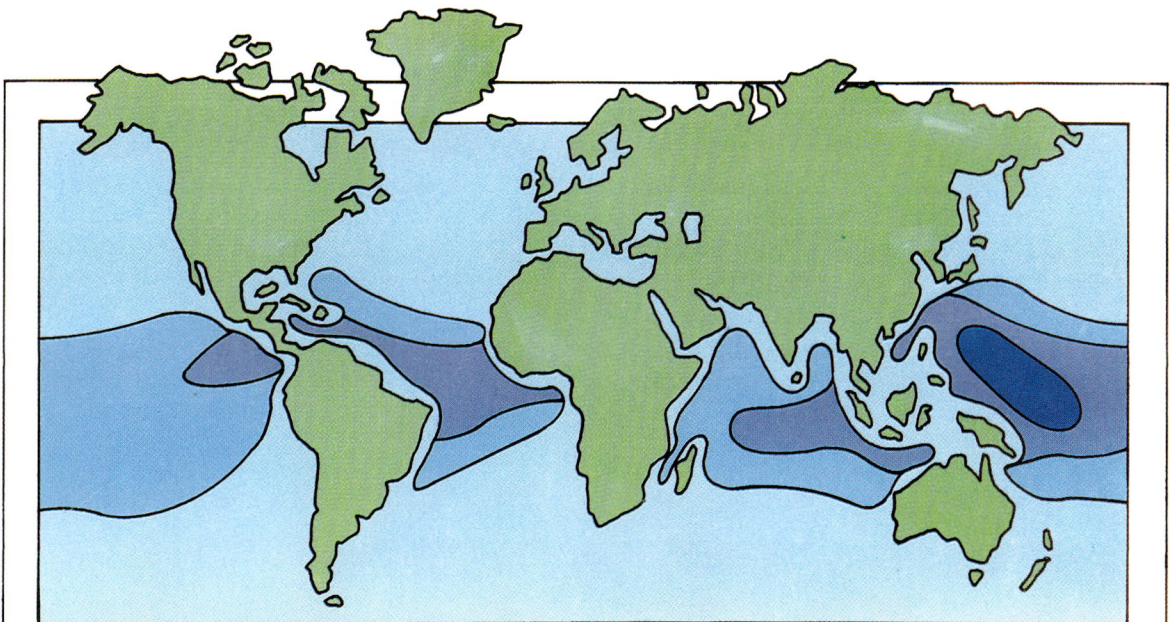

The map above shows the ocean areas where OTEC could operate efficiently.

The darkest areas show where the highest surface temperatures occur.

The fusion solution?

Without the sun, there could be no life on Earth. The sun has been burning for millions of years. The source of its energy is a reaction called *nuclear fusion*. If we could recreate that process on Earth, we would have a source of energy as limitless as the sun's.

To create fusion, temperatures of over a hundred million degrees Celsius are required. In machines like the one in the photograph, lasers are being used to try to reach these incredible temperatures. But it may be as long as 75 years before these experiments result in a working fusion power station.

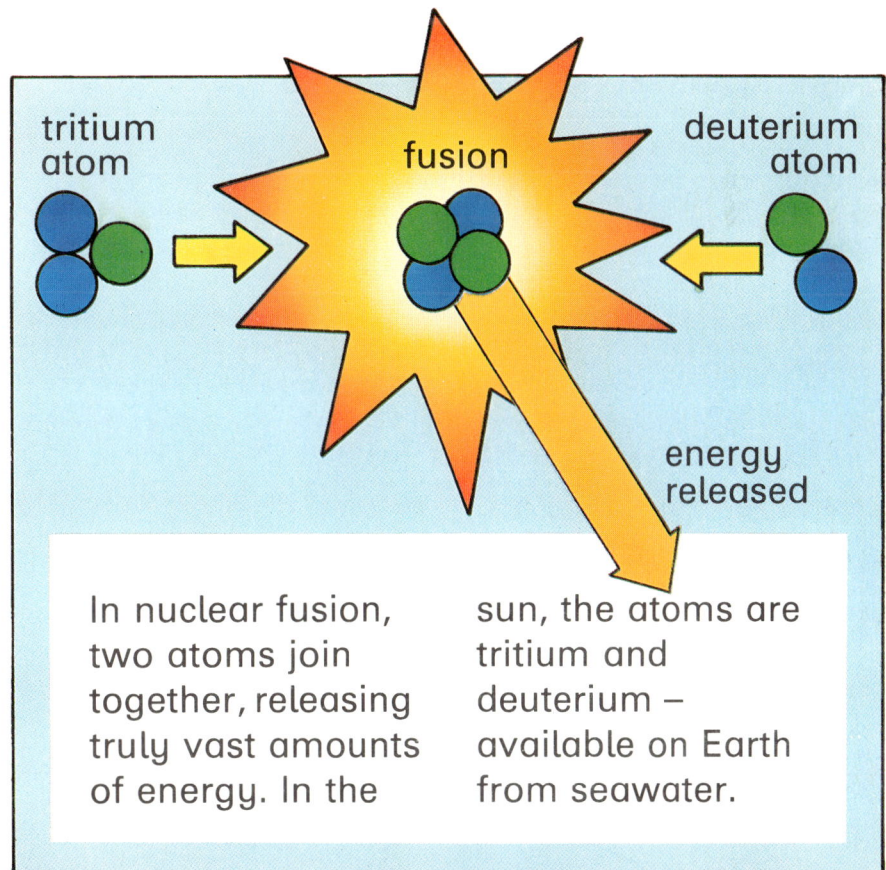

tritium atom

fusion

deuterium atom

energy released

In nuclear fusion, two atoms join together, releasing truly vast amounts of energy. In the sun, the atoms are tritium and deuterium – available on Earth from seawater.

Space power station?

In space, the sun shines 24 hours a day, and there are never any clouds to get in the way. Already, solar cells are used to power electrical systems on board satellites. They can generate electricity from sunlight which causes special chemicals to react.

American space scientists have proposed building a giant solar power station in space, many thousands of metres across. Its solar cells would generate vast amounts of power, which could be beamed down to Earth. The cost of such a project would be enormous, but once the station was working its energy would be almost free.

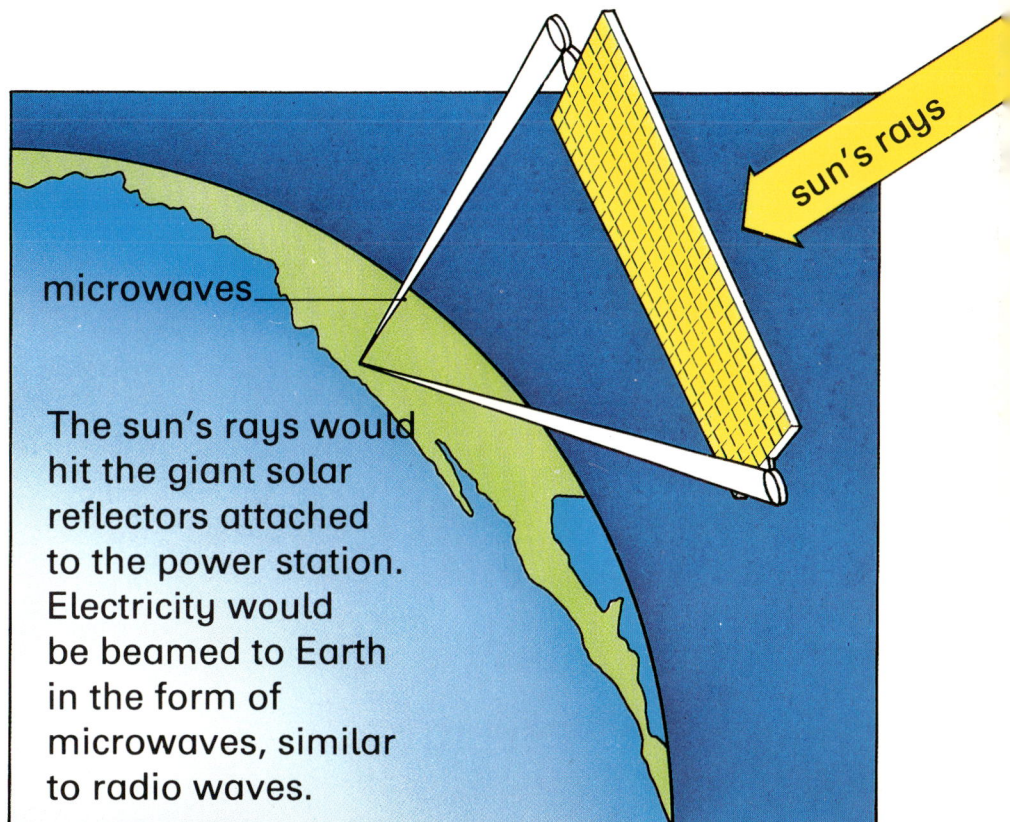

sun's rays

microwaves

The sun's rays would hit the giant solar reflectors attached to the power station. Electricity would be beamed to Earth in the form of microwaves, similar to radio waves.

Fact file 1

Coal gasification (right) is one way in which coal deposits may be extracted in the future. Poor quality coal is too expensive to mine conventionally. Fired by a stream of oxygen, the coal burns. The gas given off can be used in homes or power stations.

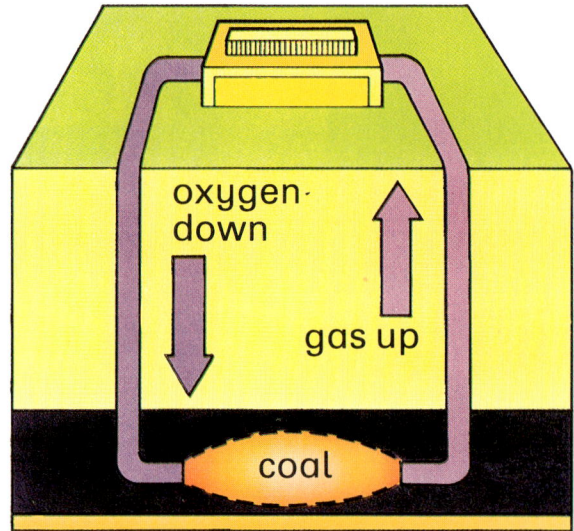

oxygen down

gas up

coal

All power stations give off most of the energy they use as heat into the atmosphere. Only just over 30% is converted to electricity. Most of the heat is lost through the cooling towers (see inset). There are projects in the USA and Europe to make use of this lost energy.

In some European countries the hot water from power stations is piped directly to homes and factories. The system, (below), called combined heat and power (CHP), doubles the amount of energy extracted from the fuel burned. A CHP plant in Denmark supplies over 150,000 homes.

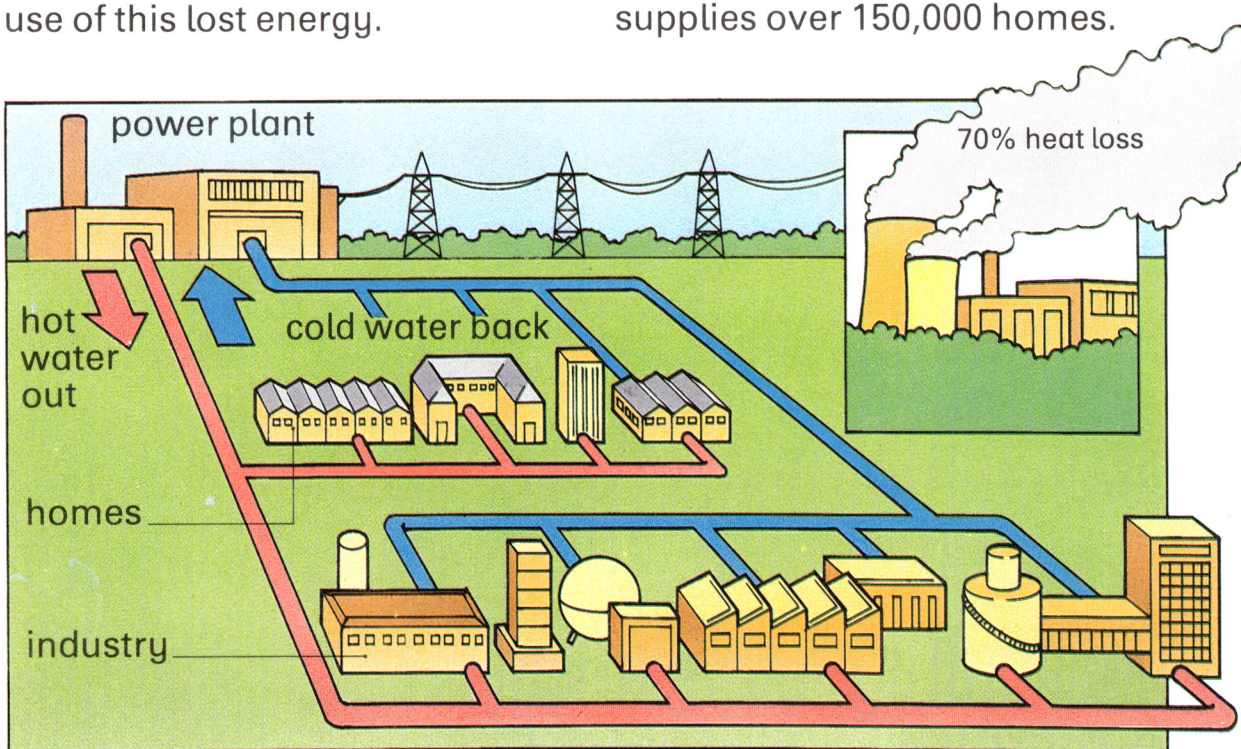

power plant

70% heat loss

hot water out

cold water back

homes

industry

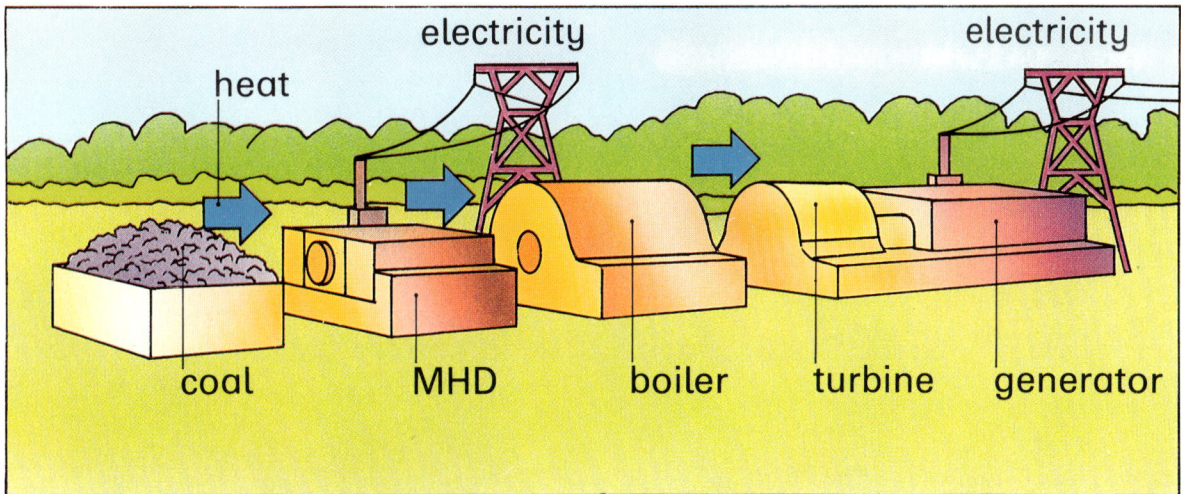

The diagram above shows another, proposed, method of getting more energy from power stations. This is called magnetohydrodynamics – MHD, for short. The idea is that the burning fuel heats gases to a very high temperature, as well as producing steam to drive turbines.

The hot gases are passed through a magnetic field, and this causes extra electricity to be generated. An MHD plant would produce twice as much electricity as an ordinary power station, from the same amount of fuel. Experimental systems have been tested in the USA and in the USSR.

Many countries have invested billions of pounds in nuclear power stations. But there has been strong opposition, both in Europe and the USA.

Nuclear power stations produce harmful wastes that can cause serious illnesses if people are exposed to them. And there is also the possibility of a serious accident – some people think the risks are too great.

Despite these fears, nuclear power is sure to provide us with energy in the future. France intends that by 1990, over 75% of its electricity will be generated by nuclear power.

Fact file 2

Different ideas for new energy sources are being developed all over the world. Some are more suitable for the richer countries. Others could help the developing countries.

In northern countries, there are large deposits of peat. The USSR burns 70 million tonnes (69 million tons) of peat per year. Peat usage is estimated to increase by fifty times by the year 2000, worldwide.

The "solar wind tower" shown below is now in operation in Spain. Warm air from the ground rises up the tower to drive a turbine, and more air is drawn in. The tower produces electricity both night and day.

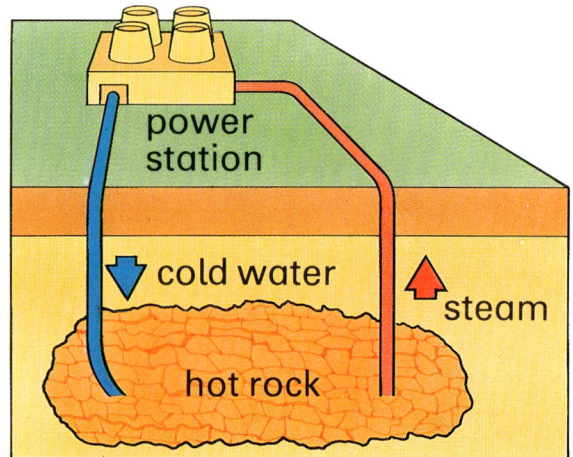

power station

cold water

steam

hot rock

Heat from the hot rocks beneath the Earth's surface is already being tapped in Iceland, New Zealand and in some Californian cities. But this geothermal power, as it is called, can also be tapped artificially (above). Cold water piped deep into the Earth is heated and forced to the surface as steam, and used to generate electricity.

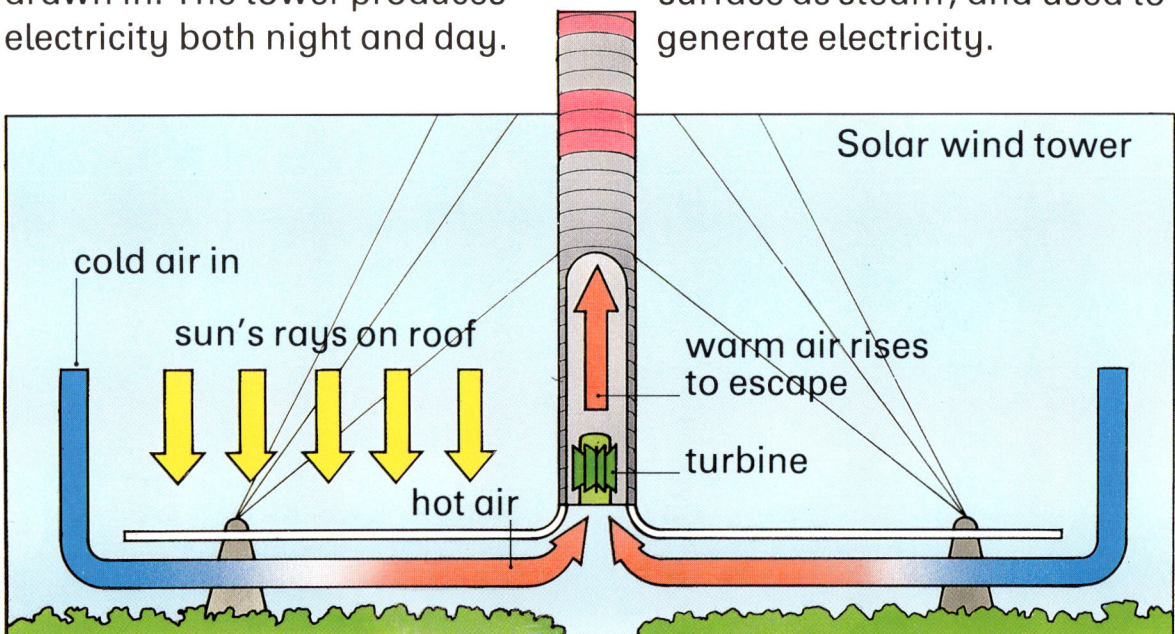

Solar wind tower

cold air in

sun's rays on roof

warm air rises to escape

turbine

hot air

Firewood comes in two forms – simple sticks and twigs, or charcoal. Charcoal burns more efficiently, but much of the wood's original energy content is lost in making it. But an open fire also loses much of its energy as heat to the air. Cheap, but efficient stoves can produce much more energy from firewood. The Lorena stove, shown opposite, is five times as efficient as an open fire.

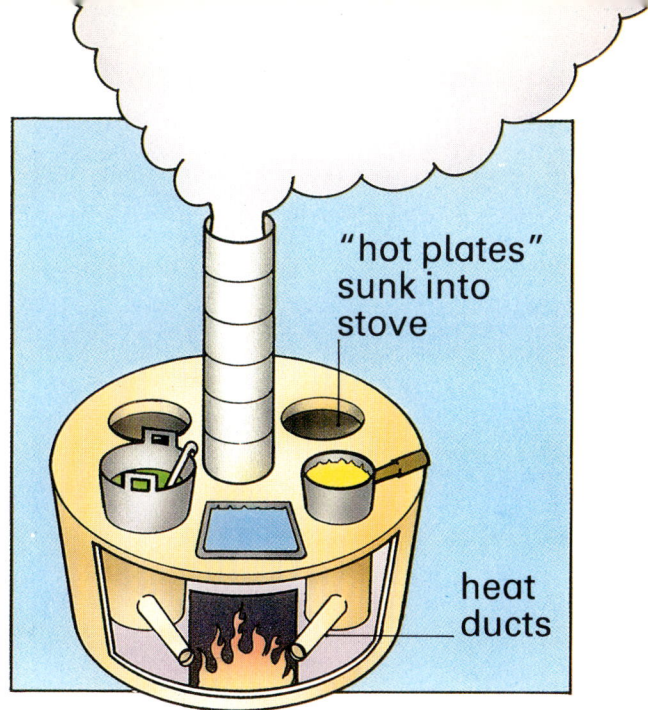

"hot plates" sunk into stove

heat ducts

Locally-available resources can provide economic fuels. For example, along the Nile in Egypt and Sudan, papyrus could be grown as a fuel crop. Countries with large livestock populations can use animal wastes. In China there are over four million "biogas" plants using animal and human wastes.

Shown below is a possible town of the future. Its energy comes from wave power, wind turbines and solar collectors (1, 2, 3). Solar power also gives domestic hot water (4). Surplus energy is stored in insulated hot water ponds (5). Solar-powered cars and even aircraft (6, 7) reduce the need for oil-based fuels.

Glossary

Fuel cell   A device which creates electricity using oxygen and hydrogen gases. So far, only small fuel cells have been made.

Geothermal power   Power obtained from the hot rocks that lie beneath the Earth's crust.

Methane   A gas that can be obtained from coal or from decaying animal and vegetable material. Biogas is in fact methane.

Nuclear fusion   A reaction in which two atoms fuse together. When this happens large quantities of energy are released. The fusion reaction creates the energy of the sun.

Nuclear power station   A type of power station that uses uranium for fuel. The atoms of uranium break up and release energy.

Recycling   The re-use of materials that would otherwise be thrown away and wasted. Much paper and many metals can be recycled.

Solar cell   A device which converts the energy of sunlight into electricity.

Turbine   A machine that rotates under steam or other pressure, to power a generator that produces electricity.

Making the Modern World
Europe

Hitler

L. E. Snellgrove

Longman

Death in the Wolf's Lair

The Wolf's Lair after the explosion

It is a very hot day in July 1944. Twenty-four German officers are sitting at a table in a wooden hut. With them is Adolf Hitler. This room is part of his headquarters near Berlin—his Wolf's Lair, as he calls it. Around him are generals and other experts helping to plan the war against Russia, Britain and the United States.

One officer, Colonel Stauffenberg, is quite near Hitler. After a time he gets up and says to Colonel Brandt beside him, 'I must go and telephone. Keep an eye on my brief case. It has secret papers in it.' The case is on the floor, resting against a chair leg. It is only six feet from Hitler. Stauffenberg walks out. Brandt leans forward to look at a map on the table. His foot knocks against the case. He picks it up and puts it on the other side of a wooden post. This post is now between Hitler and the case. By moving it nearer to himself, Brandt has made certain that he will die because inside the case is a two-pound bomb, fused and ready to go off. The minutes tick away. A general is reading a report aloud. Inside the case, acid is eating through the thin wire of the fuse.

Ten minutes pass. One or two of the generals begin to wonder why Stauffenberg is so long at the phone. Suddenly there is a terrific explosion. Flashes of yellow flame and thick black

2